Vikramarajan Jambulingam

Particle swarm optimizer

Economic dispatch with valve point effect using various PSO techniques

Anchor Compact

Jambulingam, Vikramarajan: Particle swarm optimizer: Economic dispatch with valve point effect using various PSO techniques. Hamburg, Anchor Academic Publishing 2014
Original title of the thesis: ECONOMIC DISPATCH WITH VALVE POINT EFFECT USING VARIOUS PSO TECHNIQUES

Buch-ISBN: 978-3-95489-283-9
PDF-eBook-ISBN: 978-3-95489-783-4
Druck/Herstellung: Anchor Academic Publishing, Hamburg, 2014

Bibliografische Information der Deutschen Nationalbibliothek:
Die Deutsche Nationalbibliothek verzeichnet diese Publikation in der Deutschen Nationalbibliografie; detaillierte bibliografische Daten sind im Internet über http://dnb.d-nb.de abrufbar

Bibliographical Information of the German National Library:
The German National Library lists this publication in the German National Bibliography. Detailed bibliographic data can be found at: http://dnb.d-nb.de

All rights reserved. This publication may not be reproduced, stored in a retrieval system or transmitted, in any form or by any means, electronic, mechanical, photocopying, recording or otherwise, without the prior permission of the publishers.

Das Werk einschließlich aller seiner Teile ist urheberrechtlich geschützt. Jede Verwertung außerhalb der Grenzen des Urheberrechtsgesetzes ist ohne Zustimmung des Verlages unzulässig und strafbar. Dies gilt insbesondere für Vervielfältigungen, Übersetzungen, Mikroverfilmungen und die Einspeicherung und Bearbeitung in elektronischen Systemen.

Die Wiedergabe von Gebrauchsnamen, Handelsnamen, Warenbezeichnungen usw. in diesem Werk berechtigt auch ohne besondere Kennzeichnung nicht zu der Annahme, dass solche Namen im Sinne der Warenzeichen- und Markenschutz-Gesetzgebung als frei zu betrachten wären und daher von jedermann benutzt werden dürften.

Die Informationen in diesem Werk wurden mit Sorgfalt erarbeitet. Dennoch können Fehler nicht vollständig ausgeschlossen werden und die Diplomica Verlag GmbH, die Autoren oder Übersetzer übernehmen keine juristische Verantwortung oder irgendeine Haftung für evtl. verbliebene fehlerhafte Angaben und deren Folgen.

Alle Rechte vorbehalten

© Anchor Academic Publishing, ein Imprint der Diplomica® Verlag GmbH
http://www.diplom.de, Hamburg 2014
Printed in Germany

TABLE OF CONTENTS

CHAPTER NO.	TITLES	PAGE NO.
	LIST OF FIGURE	3
	LIST OF TABLE	4
	ABSTRACT	5
1.	CHAPTER-1	7
	1.1. Introduction	7
	1.2. Methodology in brief	8
	1.3. Organization of the report	9
2.	CHAPTER-2	10
	2.1. Introduction to Economic Dispatch	10
	2.1.1. Generator Operating Cost	11
	2.2. Mathematical Analysis	11
	2.2.1. Analytical Method	14
	2.2.2. Gradient Method	14
	2.3. Valve Point Loading	14
	2.4. Problem Formulation	15
3.	CHAPTER-3	16
	3.1. Evolutionary Algorithm	16
	3.2. Anti colony Optimization	17
	3.3. Particle Swarm Optimization	17
	3.4. Overview of Particle Swarm Optimization	19
	3.5. Implementation of PSO methods in Economic Dispatch	20
	3.5.1. Advantages of PSO	21
4.	CHAPTER-4	22
	4.1. Introduction to various PSO techniques	22
	4.2. Adaptive Particle Swarm Optimization	22
	4.2.1. The procedure of Adaptive PSO	23
	4.3. Chaotic Particle Swarm Optimization	24
	4.3.1. CPSO methods for Economic Dispatch	26
	4.3.1.1. Representation of individual string	26
	4.3.1.2. Evaluation function	27
	4.3.1.3. Procedures of CPSO methods for Economic Dispatch	27
	4.4. New Particle Swarm Optimization	29
5.	CHAPTER-5	31
	5.1. Introduction	31
	5.1.1. Organization of the result	31
	5.2. The Test bus system in detail	32

		5.3. Results obtained by using the PSO	32
		5.3.1. Parameters	32
		5.3.2. Overall Report	33
		5.4. Result obtained by using the APSO	34
		5.4.1. Parameters	34
		5.4.2. Overall Report	34
		5.5. Result obtained by using the CPSO	36
		5.5.1. Parameters	36
		5.5.2. Overall Report	36
		5.6. Result obtained by using the NPSO	38
		5.6.1. Parametrs	38
		5.6.2. Overall Report	38
		5.7. Analysis of four PSO techniques	40
		5.8. Comparision of graphs	40
6.		**CHAPTER-6**	41
		6.1. Analysis of different PSO techniques	41
		6.2. Conclusion	41
		SAMPLE PROGRAMS	42
		[I] PARTICLE SWARM OPTIMIZER	42
		[II] ADAPTIVE PARTICLE SWARM OPTIMIZER	46
		[III] CHAOTIC PARTICLE SWARM OPTIMIZER	50
		[IV] NEW PARTICLE SWARM OPTIMIZER	54
		REFERENCES	58

LIST OF FIGURE

FIGURE NO.	TITLE	PAGE NO.
5.3.2.(b)	A graph of Gbest versus iteration for best solution of PSO	33
5.4.2.(b)	A graph of Gbest versus iteration for best solution of APSO	35
5.5.2.(b)	A graph of Gbest versus iteration for best solution of CPSO	37
5.6.2.(b)	A graph of Gbest versus iteration for best solution of NPSO	39
5.8.(b)	Comparison of all four PSO techniques graph	40

LIST OF TABLES

TABLE NO.	TITLE	PAGE NO.
5.2.(a)	Input data for IEEE 13 generator system	32
5.3.2.(a)	Result obtained by using PSO techniques	33
5.4.2.(a)	Result obtained by using APSO techniques	34
5.5.2.(a)	Result obtained by using CPSO techniques	36
5.6.2.(a)	Result obtained by using NPSO techniques	38
5.7.(a)	Comparison of all four PSO techniques	40

ABSTRACT

Four modified versions of particle swarm optimizer (PSO) have been applied to the economic power dispatch with valve-point effects. In order to obtain the optimal solution, traditional PSO search a new position around the current position. The proposed strategies which explore the vicinity of particle's best position found so as far leads to a better result. In addition, to deal with the equality constraint of the economic dispatch problems, a simple mechanism is also devised that the difference of the demanded load and total generating power is evenly shared among units except the one reaching its generating limit. To show their capability, the proposed algorithms are applied to thirteen.Comparision among particle swarm optimization is given. The results show that the proposed algorithms indeed produce more optimal solutions in both cases.

The different PSO techniques are New PSO,Self Adaptive PSO and Chaotic PSO.Among the different PSO techniques, it is found that Self-Adaptive PSO is better than other PSO techniques in terms of better solutions, speed of convergence, time of execution and robustness but it has more premature convergence.

CHAPTER-1

1.1 Introduction

The main purpose of Economic Load Dispatch is to minimize the total generation cost of the plant by considering the generator limits. In power generation fuel cost plays major role. Factors which influence power generation at minimum cost are operating efficiencies of generator, fuel cost and transmission losses.

Efficient generator in the system does not generate minimum cost as it may be located in an area where fuel cost is high. If the plant is located far from the load centre, transmission losses may be higher and the plant may be uneconomical.

The main aim is to identify the generation of different plants, such that the total operating cost is minimum. The major component of generator operating cost is the fuel input/hour and the maintenance cost contributes very less.

Total operating cost includes the fuel cost, cost of labour, maintenance. These costs are assumed to be a fixed percentage of the fuel cost. After neglecting the transmission losses in economic load dispatch we are considering only the generator units but not as the system.

We are neglecting the transmission line losses, line impedance etc., for analysis system is having only one bus with all generations and load are connected. As there are no transmission losses, the total load demand (Pd) is the sum of all generations.

1.2 Methodology in Brief

The coding of the algorithms was done on MATLAB 6.5, and the test system is the IEEE 13 generator system. Each algorithm was run for specified number of iterations and the best value obtained was recorded, along with the graph for the average and minimum value against the number of iterations.

The time of execution for all four algorithms were measured and recorded. Each algorithm was executed ten times and the best and the worst value were found, the graph for these executions were plotted.

The optimization techniques used are PSO, CPSO, NPSO and APSO, a random population is initialized and the fitness value of each is calculated. This population is sent through a selection process where the probability of the member of the population being selected into the matting population is directly proportional to the previously measured fitness.

Velocity limits of the generators are initialized and it has been carried out by initializing the generating velocities, besides those iterations are started and gbest values are found out by continuously updating the population and also finding the fitness of the present population.

1.3 Organization of the Report

[1] Chapter -2 contains the basic theory on Economic Load Dispatch along with some elementary mathematical background.

[2] In Chapter-3 various available algorithms are discussed.

[3] Chapter-4 explains in detail the algorithm used.

[4] Chapter-5 and Chapter-6 are result and conclusion.

CHAPTER-2

2.1 Introduction to Economic Load Dispatch

Scarcity of energy resources, increasing power generation cost and ever-growing demand for electric energy necessitates optimal economic dispatch in today's power systems. Optimal system operation involves consideration of economy of operation, system security, emission at fossil fuel plants, and optimal release of water at hydro generation.

Economic dispatch problem is to minimize the total cost of generating real power (production cost) at various stations while satisfying the loads and losses in the transmission lines. In load flow problems, two variables are specified at each bus and solutions is obtained for the variables.

In a practical power system, power plants are not loaded at the same distance from the center of loads and their fuel cost is different. The generation capacity is more than the demand and losses. So there is a need to schedule the generation. In an interconnected power system, the objective is to find the real and reactive power scheduling of each power plant in such a way to minimize the operation cost. The generators real and reactive powers are allowed to vary within certain limits to meet a particular load demand with minimum fuel cost.

Electrical energy can not be stored, but is generated from natural sources and delivered as demand arises. A transmission system is used for the delivery of bulk power over considerable distance and a distribution system is used for local deliveries. An interconnected power system consists of mainly three parts :

        1. The generator, which produce electrical energy

        2. The transmission line which transmit it to far away places

        3. The load which use it

Such a configuration applies to all inter connected networks, where the number elements may vary. The transmission networks are interconnected through ties so that utilities can exchange power, share reserves and render assistance to one another in times of need. Since the sources of energy are so diverse , the choice of one or the other is made on economic, technical or geographic basic. As there are few facilities to store electric energy, the net production of utility must clearly track its total load for an inter connected system, the fundamental problem is one of minimizing the source expenses. The economic dispatch problem is to define the production level of each plant so that the total cost of generation and transmission for a prescribed shecdule of loads.

- Forecasting includes determining the peak rate of supply i.e, energy demand for both long-term investment decisions and short-term operating decisions.
- Operating applications include allocation of out put, unit start-up selection, hydro thermal co ordinations and maintenance scheduling.
- The investment planning applications cover the generation and transmission system.

2.1.1 Generator operating cost:

Factors which influence power generation at minimum cost are operating efficiencies of generator, fuel cost and transmission losses.

Efficient generator in the system does not generate minimum cost as it may be located in an area where fuel cost is high. If the plant is located far from the load centre, transmission losses may be higher and the plant may be uneconomical.

The main aim is to identify the generation of different plants, such that the total operating cost is minimum. The major component of generator operating cost is the fuel input/hour and the maintenance cost contributes very less.

2.2 Mathematical Analysis

He we are considering only the generating units, but not as the system. We are neglecting the transmission line losses, line impedance etc. for analysis, the system is having only one bus with all generations and load are connected.

As there is no transmission losses, the total demand P_D is the sum of all generation. For each plant assume the cost function FC_i

$$\text{Min } FC_{\text{total}} = \sum_{i=1}^{N_G} FC_i \quad \text{-----(1)}$$

$$= \sum_{i=1}^{N_G} \alpha_i + \beta_i P_i + \gamma_i P_i^2 \quad \text{-----(2)}$$

Subject to the constraint,

$$\sum_{i=1}^{N_G} P_i = P_D \quad \text{-----(3)}$$

The power output of any generator should not exceed the its rating nor should it be below that necessary for stable turbine operation thus, the generations are restricted to lie within given minimum and maximum limits. The problem is to find the real power generation for each plant such that the objective function as defined by (2) is minimum, subject to the constrain given by (3) and the inequality constraints given by,

$P_i^{min} \leq P_i \leq P_i^{max}$ where $i = 1,2,3,\ldots\ldots\ldots N_G$

P_i^{min}, P_i^{max} – minimum and maximum generating limits

FCtotal – total production cost

FCi – production cost of ith plant

Pi – power generation of ith plant

PD – total load demand

NG – total number of generating unit

Using Lagrange Multipliers,

$$L = FC_{total} + \lambda (P_D - \sum_{i=1}^{N_G} P_i)$$

The minimum value will be obtained at the poing where the partials of the function to its variables are zero.

$$\frac{\partial L}{\partial P_i} = 0$$

$$\frac{\partial L}{\partial \lambda} = 0$$

$$\frac{\partial L}{\partial P_i} = \frac{\partial FC_{total}}{\partial P_i} + \lambda (0 - 1) = 0$$

$$\frac{\partial FC_{total}}{\partial P_i} + \lambda (0 - 1) = 0$$

$$FC_{total} = FC_1 + FC_2 + \ldots\ldots\ldots FC_{N_G}$$

$$\frac{\partial FC_{total}}{\partial P_i} = \frac{dFC_i}{dP_i}$$

Therefore optimal dispatch condition is,

$$\frac{dFC_i}{dP_i} = \lambda$$

$$\frac{dFC_i}{dP_i} = \beta_i + 2\alpha_i P_i$$

$$\lambda = \beta_i + 2\alpha_i P_i$$

$$\frac{\partial L}{\partial \lambda} = P_D - \sum_{i=1}^{N_G} P_i = 0$$

$$\sum_{i=1}^{N_G} P_i = P_D$$

When losses are neglected, for most economic operation all plants must operate at equal incremental production cost.

$$P_i = \frac{\lambda - \beta_i}{2\gamma_i}$$

This is the coordination equation which is a function of λ. So,

$$\sum_{i=1}^{N_G} \frac{\lambda - \beta_i}{2\gamma_i} = P_D$$

$$\lambda = \frac{P_D + \sum_{i=1}^{N_G} \frac{\beta_i}{2\gamma_i}}{\sum_{i=1}^{N_G} \frac{1}{2\gamma_i}}$$

The value of λ has to be substituted in

$$P_i = \frac{\lambda - \beta_i}{2\gamma_i}$$, To obtain the optimal scheduling of generation.

To get the economical values of Pi, it has to undergo iterative process. Using gradient method, we get the solutions quickly.

$$f(\lambda) = P_D$$

Expanding the left hand side in Taylors Series above an operating point

$$f(\lambda)^{(k)} + (\frac{df(\lambda)}{d\lambda}) \Delta \lambda^{(k)} = P_D$$

$$\Delta \lambda^{(k)} = \frac{\Delta P^{(k)}}{(\frac{df(\lambda)}{d\lambda})^{(k)}}$$

$$\Delta P^{(k)} = P_D - \sum P_i^{(k)}$$

$$= \frac{\Delta P^{(k)}}{\sum (\frac{dP_i}{d\lambda})^{(k)}}$$

$$\Delta \lambda^{(k)} = \frac{\Delta P^{(k)}}{\sum_{i=1}^{N_G} \frac{1}{2\gamma_i}}$$

$$\lambda^{(k+1)} = \lambda^{(k)} + \Delta \lambda^{(k)}$$

Economic load dispatch problems can be solved theoretically by the following two methods they are as follows:

1. Analytical method
2. Gradient method

2.2.1 Analytical method

In this method the λ is determined by solving the given parameters

Where,
$$\lambda = \frac{P_D + \sum_{i=1}^{N_G} \frac{\beta_i}{2\gamma_i}}{\sum_{i=1}^{N_G} \frac{1}{2\gamma_i}}$$

α, β, γ – Cost coefficients
i – Index of the generator
P_D – total load demand
N_G – total number of generating units

2.2.2 Gradient method

In this method λ value is assumed (λ= 0 to 1)

2.3 Valve Point Loading

When the load demand increases the speed of the generator will decrease automatically. We know that generator is coupled with prime mover (turbine), so to increase the speed of the generator the valve point connected in the turbine is opened gradually, then the turbine starts rotating faster hence generator gets back almost to the its original speed.

Valve point effect is added in the economic load dispatch problem to increase the accuracy of the total fuel cost however the cost function of a generator is not always differentiable due to the valve-point effects and/or change of fuels. The valve-point effects introduce ripples in the heat-rate curve. The fuel cost function with valve-point loadings of the generators is usually modeled as

$$FC_i = \sum_{i=1}^{N_G} a_i + b_i P_i + c_i P_i^2 + \left| e_i \sin(f_i(P_{i(min)} - P_i)) \right|$$

Where, a_i, b_i, c_i are fuel cost coefficients of generator i

e_i and f_i are fuel cost coefficients of generator i with valve point effect.

2.4 Problem Formulation

The economic dispatch problem is formulated as,

$$\text{Min } FC_{total} = \sum_{i=1}^{N_G} FC_i$$

$$\text{Subject to, } \sum_{i=1}^{N_G} P_i = P_D$$

$$P_i^{min} \leq P_i \leq P_i^{max} \quad \text{Where } i = 1, 2, 3, \ldots \ldots N_G$$

CHAPTER-3

3.1 Evolutionary Algorithm

An evolutionary algorithm (EA) is the subset of evolutionary computation, a generic population – based metaheuristic optimization algorithm. An EA uses some mechanism inspired by biological evolution: reproduction, mutation, recombination, natural selection and survival of the fittest. Candidate solutions to the optimization problem play the role of individuals in a population, and the cost function determines the environment within which the solution "live (see also fitness function)". Evolution of the population then takes place after the repeated application of the above operators. Artificial evolution (AE) describes a process involving individual evolutionary algorithm; EAs are individual components that participate in artificial evolutions.

EAs consistently perform well approximating solutions to all types of problems because they don't make any assumption about the underlying fitness landscape; this generality is show by successes in fields as diverse as engineering, art, biology, economics, genetic, operations research, robotics, social sciences, physics and chemistry. However, evolutionary algorithms can nonetheless the outperformed by more field – specific algorithm.

Apart from their use as mathematical optimizers, evolutionary computation and algorithms and been used as an experimental frame work within which to validate theories about biological evolution and natural selection, particularly through work in field of artificial life. Techniques from evolutionary algorithm applied to the modeling of biological evolution are generally limited to explorations of micro evolutionary processes. A limitation of evolutionary algorithms is their lack of clear genotype – phenotype distinction. In nature, the fertilized egg cell undergoes a complex process know as embryogenesis to become a mature phenotype. This indirect encoding is believed to make genetic search more robust (i.e. reduce the probability of fatal mutations), and also may improve the resolvability of the organism. Recent work in the field of artificial embryogeny, or artificial developmental systems, seeks to address these concerns.

Four evolutionary methods are used in this project they are as follows PSO, APSO, CPSO and NPSO, these algorithms are discussed in detail.

3.2 Ant Colony Optimization

In the real world, ants (initially) wander randomly, and upon finding food return to their colony while laying down phenomenon trails. If other ants find such a path, they are not likely not to keep travelling at random, but to instead follow the trail, returning and reinforcing it if they eventually find the source of food.

Over time, however the phenomenon trails starts to evaporate thus reducing the attractive strength. The more time it takes from an ant to travel down the path and back again, the more time the phenomenons have to evaporate. A short path by comparison, gets marched over faster, and thus the phenomenon density remains high as it is laid on the path as fast as it can evaporate. Phenomenon evaporation has also the advantage of avoiding the convergence to a locally optimal solution. If there were no evaporation at all, the path chosen by the first ants would tend to be excessively attractive to the following ones. In that case, the exploration of the solution space would be constrained.

Thus, one ant finds a good (short, in other words) path from the colony to a food source, other ants are more likely to follow that path, and positive feedback algorithm is to mimic this behavior with "simulated ants" walking around the graph representing the problem to solve.

Ant colony optimization algorithms have been used to produce near – optimal solutions to the travelling salesman problem. They have an advantage over simulated annealing and genetic algorithm approaches when the graph may change dynamically; the ant colony algorithm can be run continuously and adapt to changes in real time. This is of interest in network routing and urban transportation systems.

3.3 Particle Swarm Optimization

Kennedy and Eberhart first introduced PSO in year 1995. The features of the method are as follows:

The method is based on researches about swarms such as fish schooling and a flock of birds. It is based on a simple concept. Therefore, the computation time is short and it requires less memory. It was originally developed for nonlinear optimization problems with continuous variables. However, it is easily expanded to treat problems with discrete variables. Therefore, it is applicable to both continuous and discrete variables. The basic assumption behind the PSO algorithm is, birds find food by flocking and not individually. This leads to the assumption that information is owned jointly in flocking.

Particle swarm optimization (PSO) is a form of swarm intelligence. Imagine a swarm of insects or a school of fish. If one sees a desirable path to go (e.g., for food, protection, etc.) the rest of swarm will be able to follow quickly even if they are on the opposite side of the swarm. On the other hand, in order to facilitate felicitous exploration of the search space, typically one wants to have each particle to have a certain level of "craziness" or randomness in their movement.

This is modeled by particles in multidimensional space that have a position and a velocity. These particles are flying through hyperspace (i.e. Rn) and have two essential reasoning capabilities: their memory of their own best position and knowledge of the swarms best. Members of the swarm communicate good positions to each other and adjust their own position and velocity based on these good positions. There are two main ways this communication is done.
- "A global" best that is known to all.
- "neighborhood" bests where each particle only communicates with a subset of the swarm about the best positions.

An algorithm is presented below where there is a global best rather than neighborhood bests. Neighborhood bests allow better exploration of the search space and reduce the susceptibility of pso to falling into local minima, but he slow down convergence speed. Note that neighborhoods merely slow down the proliferation of new best, rather than creating isolated subswarms because of the overlapping of neighborhoods: to make neighborhoods of size 3, say, particle 1 would only communicate with particles 2 through 5, particle 2 with 3 through 6, and so on. But then a new best position discovered by particle 2's neighborhood would be communicated to particle 1's neighborhood at the next iteration of the pso algorithm presented below.

Smaller neighborhoods to faster convergence, with a global best representing a neighborhood consisting of the entire swarm.

In the following sections we will deal with the main algorithms being used in the program.

3.4 Over view of Particle Swarm Optimization

Basically PSO was developed for two-dimension solution space by Kennedy and Eberhart. The position of each individual is represented by XY axis position and its velocity is expressed by V_x in x direction and V_y in y direction. Modification of the individual position is realized by the velocity and position information.

PSO algorithm for N-dimensional problem formulation based on the above concept can be described as follows. Let P be the 'particle' co-ordinates (position) and V its speed (velocity) in a search space. Consider i as a particle in the total population (swarm). Now the ith particle position can be represented as,

$P_i = (P_{i1}, P_{i2}, P_{i3}, \ldots P_{iN})$ in the N-dimensional space.

The best previous position of the i_{th} particle is stored and represented as,

$P_{besti} = (P_{besti1}, P_{besti2}, \ldots P_{bestij})$.

All the P_{best} are evaluated by using a fitness function, which differs for different problems. The best particle among all P_{best} is represented as g_{best}. The velocity of the i_{th} particle is represented as,

$V_i = (V_{i1}, V_{i2}, \ldots V_{ij})$

The modified velocity of each particle can be calculated using the information, (i) the current velocity (ii) the distance between the current position and Pbest and (iii) the distance between the current position and gbest. This can be formulated as an equation,

$$v_i^{k+1} = w_i v_i^k + (c_x * rand * (pbest - s_i^k)) + (c_2 * rand * (gbest - s_i^k)) \quad \text{-----(4)}$$

The current position can be modified by the following equation:

$$P_i^{(k+1)} = P_i^k + v_i^{k+1}$$

v_i^k -Current velocity of particle i at iteration k

v_i^{k+1} -Modified velocity of particle i

P_i^k -Current position of particle i at iteration k

rand -Random number between 0 and 1

S_i^k -Current position of particle i at iteration k

$pbest_i$ -pbest of particle i

gbest -gbest of the group

w -inertia weight factor

c1, c2 -acceleration constant

The use of linearly decreasing inertia weight factor w has provided improved performance in all the applications. Its value is decreased linearly from about 0.9 to 0.4 during a run. Suitable selection of the inertia weight provides a balance between global and local exploration and exploitation, and results in fewer iterations on average to find a sufficiently optimal solution. Its value is set according to the following equation,

$$w = w_{max} - \frac{w_{max} - w_{min}}{iter_{max}} * iter \quad \text{-----(5)}$$

Where, w_{max} and w_{min} are both random numbers called initial weight and final weight

$iter_{max}$ - the maximum iteration number

iter - the current iteration number

$w_{max} = 0.9$

$w_{min} = 0.4$

and $C_1 = C_2 = 2.0$.

In Eq. (4) the first term indicates the current velocity of the particle, second term represents the cognitive part of PSO where the particle changes its velocity based on its own thinking and memory. The third term represents the social part of PSO where the particle changes its velocity based on the social-psychological adaptation of knowledge.

3.5 Implementation of PSO method in ED

Implementation of PSO in Economic Dispatch involves the following steps:

Step (1) In the ED problems the number of on-line generating units is the 'dimension' of this problem. The particles are randomly generated between the maximum and the minimum operating limits of the generators.

For example, if there are N units, the i_{th} particle is represented as follows,

$P_i = (P_{i1}, P_{i2}............P_{iN})$

Step (2) The particle velocities are generated randomly in the range $[-V_i^{max}, V_i^{max}]$.

The maximum velocity limit in the ith dimension is,

$$V_i^{max} = \frac{P_{i,max} - P_{i,min}}{R}$$

Where, R is the chosen number of intervals in the ith dimension.

Step (3) Objective function values of the particles are evaluated using the respective objective functions given by equ. (1), (2). These values are set as the Pbest value of the particles.

Step (4) The best value among all the Pbest values, gbest, is identified.

Step (5) New velocities for all the dimensions in each particle are calculated using equ. (4).

Step (6) The position of each particle is updated using equ. (5).

Step (7) The objective function values are calculated for the updated positions of the particles If the new value is better thant the previous Pbest, the new value is set to Pbest. If the stopping criteria are met, the positions of particles represented by gbest are the optimal solutions. Otherwise, the procedure is reapeated from step (4).

3.5.1 Advantages of PSO

(1) PSO is easy to implement, and there are few parameters to adjust.

(2) Unlike the GA, PSO has no evolution operators such as crossover and mutation.

(3) In GAs, chromosomes share information so that the whole population moves like one group, but in PSO, only Global best particle (Gbest) gives out information to the others. It is more robust.

(4) PSO can be more efficient than GAs; that is, PSO often finds the solutions with fewer objective function evaluations than are required by GAs.

(5) PSO uses payoff (performance index or objective function) information to guide the search in the problem space.

(6) Unlike GAs and other heuristic algorithms, PSO has the flexibility to control the balance between global and local exploration of the search space. This unique feature of PSO overcomes the premature convergence problem and enhances the searchces the search capability.

CHAPTER-4

4.1 Introduction to various PSO techniques

There are different types of PSO techniques where the few techniques used here are APSO, CPSO and NPSO.

4.2. Adaptive Particle Swarm Optimization

In the simple PSO method, the inertia weight is made constant for all the particles in a single generation, but the most important parameter that moves the current position towards the optimum position is the inertia weight (x). In order to increase the search ability, the algorithm should be redefined in the manner that the movement of the swarm should be controlled by the objective function. In our adaptive PSO, the particle position is adjusted such that the highly fitted particle (best particle) moves slowly when compared to the lowly fitted particle. This can be achieved by selecting different x values for each particle according to their rank, between x_{min} and x_{max} as in the following form,

$$w_i = w_{min} + \frac{(w_{max} - w_{min}) * Rank_i}{Total population} \quad \text{-----(6)}$$

So, from Eq. (6), it can be seen that the best particle takes the first rank, and the inertia weight for that particle is set to the minimum value while that for the lowest fitted particle takes the maximum inertia weight, which makes that particle move with a high velocity. The velocity of each particle is updated using Eq. (7), and if any updated velocity goes beyond V_{max}, it is limited to V_{max} using Eq. (8).

$$v_{ij}(t+1) = w_i v_{ij}(t) + c_1 r_1 (p_{ij}(t) - x_{ij}(t)) + c_2 r_2 (p_{gj}(t) - x_{ij}(t)) \quad \text{-----(7)}$$

$$v_{ij}(t+1) = sign(v_{ij}(t+1)) * min(|v_{ij}(t+1)|, v_{jmax}) \quad \text{-----(8)}$$

$j = 1, 2, \ldots \ldots d.$
$i = 1, 2, \ldots \ldots n.$

The new particle position is obtained by using Eq. (9), and if any particle position goes beyond the range specified, it is adjusted to its boundary using Eq. (10)

$$x_{ij}(t+1) = x_{ij}(t) + v_{ij}(t+1), \quad \text{-----（9）}$$

$j = 1, 2, \ldots\ldots\ldots d.$
$i = 1, 2, \ldots\ldots\ldots n.$

$$x_{ij}(t+1) = \min(x_{ij}(t+1), range_{j\max}),$$
$$x_{ij}(t+1) = \max(x_{ij}(t+1), range_{j\min}) \quad \text{-----（10）}$$

The concept of re-initialization is introduced to the proposed APSO algorithm after a specific number of generations if there is no improvement in the convergence of the algorithm. The population of the proposed APSO at the end of the above mentioned specific generation is re-initialized with new randomly generated individuals. The number of these new individuals is selected from the k least fit individuals of the original population, where 'k' is the percentage of the total population to be changed. This effect of population re-initialization is, in a sense, similar to the mutation operator in a GA. This effect is favorable when the algorithm prematurely converges to a local optimum and further improvement is not noticeable. This re-initialization of population is performed after checking the changes in the 'F_{best}' value in each and every specific number of generations.

4.2.1 The procedure of Adaptive PSO

Step1: Get the input parameters like range [min max] for each of the variables, c_1, c_2, Iteration counter = 0, V_{max}, w_{min} and w_{max}.

Step2: Initialize n number of population of particles of dimension di with random Positions and velocities.

Step3: Increment iteration counter by one.

Step4: Evaluate the fitness function of all particles in the population, find particles best Position Pbest of each particle and update its objective value. Similarly, find the global best position (Gbest) among all the particles and update its objective value.

Step5: If stopping criterion is met go to step (11). Otherwise continue.

Step6: Evaluate the inertia factor according to Eq. (6), so that each particles movement is directly controlled by its fitness value.

Step7: Update the velocity using Eq. (7) and correct it using Eq. (8).

Step 8: Update the position of each particle according to Eq (9), and if the new position goes out of range, set it to the boundary value using Eq. (10).

Step 9: The elites are inserted in the first position of the new population in order to maintain the best particle found so far.

Step 10: For every generations, this $F_{Best,new}$ value (at the end of these 5 generations) is compared with the $F_{Best,old}$ value (at the beginning of these 5 generations), if there is no noticeable change, then re-initialize k% of the population. Go to step (3).

Step 11: Output the Gbest particle and its objective value.

4.3. Chaotic Particle Swarm Optimization

PSO has gained much attention and widespread applications in different fields. However, the performance of the simple PSO greatly depends on its parameters, and it often suffers the problem of being trapped in local optima so as to prematurely converge [24]. In order to avoid these disadvantages, Liu et al. proposed a chaotic particle swarm optimization (CPSO) method that combines PSO with AIWF (adaptive inertia weight factor) and chaotic local search (CLS) based on the logistic equation [25].

$$w = w_{min} + \frac{(w_{max} - w_{min})(f - f_{min})}{f_{avg} - f_{min}}, \quad f \leq f_{avg}$$

$$w_{max} \quad\quad\quad\quad\quad\quad\quad\quad f \geq f_{avg}$$

Where, w_{max} and w_{min} maximum and minimum of w

f is the current objective value of the particle

f_{avg} and f_{min} are the average and minimum objective values of all particles.

The logistic equation, which exhibits sensitive dependence on initial conditions, is introduced in the process of chaotic local search defined by the following equation.

$$cx_i^{(k+1)} = 4cx_i^{(k)}(1-cx_i^{(k)}), \qquad i = 1, 2, 3............n$$

Where, cxi denotes the ith chaotic variable and k represents the iteration number. Obviously, $cx_i^{(k)}$ is distributed in the interval (0, 1.0) under the conditions that the initial $cx_i^{(0)} \in (0,1)$ and that $cx_i^{(0)} \notin \{0.025, 0.5, 0.75\}$

The procedures of CLS based on the logistic equation can be illustrated as follows.

Step 1. Setting k=0 and mapping the decision variables $x_i^{(k)}$, I = 1, 2,.................n among The intervels (xmin,I, xmax,i), I = 1,2,................n to the chaotic variables $cx_i^{(k)}$ Located in the interval (0,1) using the following equation:

$$cx_i^{(k)} = \frac{x_i^k - x_{min,i}}{x_{max,i} - x_{min,i}}, \qquad I = 1, 2,..................n$$

Step 2. Determine the chaotic variables $cx_i^{(k+1)}$ for the next iteration using the logistic equation According to $cx_i^{(k)}$

Step 3. Convert the chaotic variables $cx_i^{(k+1)}$ to the decision variables a $x_i^{(k+1)}$ using the Following equation :

$$x_i^{(k+1)} = x_{min,i} + cx_i^{(k+1)}(x_{max,i} - x_{min,i}), \qquad I = 1, 2,..................n$$

Step 4. Evalute the new solution with decision variables $x_i^{(k+1)}$, I = 1, 2,..............n

Step 5. If the new solution is better than $X_{(0)} = [[x_1^{(0)},................,x_n^{(n)}]$ or the predefined Maximum iteration is reached, output the new solution as the result of the CLS;Otherwise, let k = k+1 and go to back to step 2.

We employed the CPSO method to solve the ED problem in a power system. During this course, we further studied on this method and proposed a new chaos particle swarm optimization that combines PSO with AIWF and a new CLS based on the Tent equation.

The Tent equation is another famous chaos system. It can also be introduced to the process of the chaotic local search, which can be defined by the following equation:

$$cx_i^{(k+1)} = 2cx_i^{(k)}, \quad 0 < cx_i^{(k)} < 1/2,$$

$$2(1 - 2cx_i^{(k)}), \quad 1/2 < cx_i^{(k)} < 1,$$

Where, cx_i and k denote the same as stated before

4.3.1 CPSO methods for EP

In this paper, the process to solve ED problems with constraints using CPSO methods was developed to get efficiently a high quality solution in practical power system operation. The CPSO methods were mainly used to find the optimal generation power of each unit that was submitted to operation at the specific period, thus minimizing the total generation cost.

4.3.1.1 Representation of individual string

The generation power output of every unit was chosen as a gene, and many genes comprise an individual, which represents a candidate solution for the ED problems. For example, we suppose there are n units that should be operated to provide power to loads. Then, we define the ith individual Pgi as follows.

$$pg_i = [p_{i1}, p_{i2}, \ldots, p_{id}], \quad I = 1, 2, \ldots n$$

Where, n – size of the population
       d – generator number
       P_{id} – generation power output of the d_{th} unit

4.3.1.2 Evalution function

The fitness of each individual is evaluated by the evaluation function. All the evaluation values are normalized into the range between 0 and 1 to emphasize the "best" chromosome and speed up convergence of the iteration procedure. We choose an evaluation function like that of Ref. [13]. The evaluation function is shown as Eqs. (11)–(13), which is the reciprocal of the generation cost function and power balance constraint as in Eqs. (1) and (2). It is obvious that if the values of Fcost (Pgi) and Ppbc (Pgi) of individual Pgi were small, then its evalutation value would be large.

$$f = \frac{1}{F_{cost} + P_{pbc}} \quad \text{-----(11)}$$

$$F_{cost} = 1 + abs\frac{(\sum_{i=1}^{n} F_i(P_i) - F_{min})}{(F_{max} - F_{min})} \quad \text{-----(12)}$$

$$P_{pbc} = 1 + (\sum_{i=1}^{n} P_i - P_D - P_L)^2 \quad \text{-----(13)}$$

Where, F_{max} – maximum generation cost
F_{min} – minimum generation cost

4.3.1.3 Procedures of CPSO methods for ED

The search procedures of the proposed CPSO methods for ED are described as follows:

Step 1. Specify the lower and upper bound generation power loads of every unit and calculate the Value of Fmax and Fmin. According to the limit of each unit, including individual Dimenstions, searching point and velocities, initialize the individuals of the population Randomly. These initial individuals should be feasible candidate solutions that satisfy All the practival operation constraints.

Step 2. For each individual Pg in the population, calculate the transmission loss PL using the B Coefficient loss formula.

Step 3. Calculate the evaluation value of each individual Pgi of the population employing the Evaluation function f determined according to the practical conditions.

Step 4. Compare each individual's evaluation value with its Pbest. Make the best evaluation Value among the Pbest be Pgbest.

Step 5. Update the member velocity of each individual based on eq. (14)

$$V_{id}[t+1] = wv_i[t] + c_1 \times rand(.) \times (P_{bestid} - Pg_{id}) + c_2 \times Rand(.) \times (P_{gbestd} - Pg_{id}), \text{--(14)}$$

$i = 1, 2, \ldots \ldots n,$
$d = 1, 2, \ldots \ldots m.$

where, n – population size
        m – number of units

Step 6. If $V_{id}[t+1] > V_d^{max}$, then $V_{id}[t+1] = V_d^{max}$

If $V_{id}[t+1] < V_d^{min}$, then $V_{id}[t+1] = V_d^{min}$

Step 7. Update the member position of each individual Pgi besed on Eq. (15)

$$Pg_{ig}[t+1] = Pg_{id}[t] + V_{id}[t+1] \text{ -----(15)}$$

Step 8. If the evaluation value of each individual is better than the previous Pbest, the current Value is set to be Pbest. If the best is better than Pgbest, the value is set to be Pgbest.

Step 9. Reserve the top n/5 particles in the population.

Step 10. Implement the chaotic local search for the best particle and update the best particle Using the result of CLS with variables $P_{g,i}^{(k)}$, $I = 1, 2, \ldots \ldots n$

Step 11. If a stopping criterion is satisfied, output the solution found best so far.

Step 12. Decrease the search space:

$$P_{min\,i} = \max(P_{min\,i}, P_{g,i}^{(k)} - r(P_{max\,i} - P_{min\,i})). \quad 0 < r < 1 \text{ -----(16)}$$

$$P_{max\,i} = \min(P_{max\,i}, P_{g,i}^{(k)} - r(P_{max\,i} - P_{min\,i})). \quad 0 < r < 1 \text{ -----(17)}$$

Step 13. Generate randomly 4n/5 new particles within the decreased search space and evaluate Them.

Step 14. Reconstruct the new population consisting of the 4n/5 new particles and the old top n/5 particles in which the best particle is replaced by the results of CLS.

Step 15. If the iterations number reaches the maximum, go to step 16. Otherwis, let k=k+1 And go back to step 2.

Step 16. The individual that generates the latest Pgbest is the optimal generation power of Each unit when the system reaches the minimum total generation cost.

4.4. New Particle Swarm Optimization

PSO is a population-based, self-adaptive, stochastic optimization technique. The basic idea of the PSO is the mathematical modeling and simulation of the food searching activities of a swarm of birds (particles). In the multidimensional space where the optimal solution is sought, each particle in the swarm is moved toward the optimal point by adding a velocity with its position.

The velocity of a particle is influenced by three components, namely, inertial, cognitive, and social. The inertial component simulates the inertial behavior of the bird to fly in the previous direction. The cognitive component models the memory of the bird about its previous best position, and the social component models the memory of the bird about the best position among the particles (interaction inside the swarm). The particles move around the multidimensional search space until they find the food (optimal solution). Based on the above discussion, the mathematical model for PSO is as follows.

Velocity update equation is given by,

$$v_i^{k+1} = w_i v_i^k + (c_{1g} * r1 * (pbest_i - s_i^k)) + (c_{1w} * r2 * (s_i^k - pworst_i)) + (c_2 * r3 * (gbest_i - s_i^k))$$

r_1, r_2, r_3 are the random no. between [0,1]

pworst$_i$ --- pworst of particle i
$C_{lg} = 1.4$
$C_{lw} = 0.6$

Initialization of the Best and Worst Positions

In the strategy of PSO, the particle's best position (P_{best}) and global best position(G_{best}) are the key factors. The best position of a particle is the position, which gives the minimum PF_T, and the best position out of all the P_{best} is taken as. In this paper, the particle's worst position (P_{worst}) is introduced. At the beginning of the iteration process, the P_{best} and P_{worst} for all the particles are taken as the same as the initial positions. The PF_T at G_{best}, is taken as $F_{Gbest}{}^0$

The search procedures of the proposed NPSO methods for ED are described as follows:

Step 1 : Input data's are to be given.

Step 2 : Initialization of position, velocity, Pbest, Pworst, Gbest and iteration count.

Step 3 : Incrasing the iteration count.

Step 4 : Updating the position, velocity, Gbest,Pbest and Pworst.

Step 5 : Invoke LRS (local random search)

Step 6 : Better Gbest optained by LRS.

Step 7 : Bettter Gbest value is not obtained by LRS

Step 8 : Repalce Gbest of NPSO with optimum of LRS

Step 9 : Maximum iteration for NPSO

Step 10: Result is obtained.

CHAPTER-5

5.1 Introduction

This chapter presents a detailed report of the results obtained for the Economic Dispatch problem, using the four optimization techniques explained in the previous chapters. The test system is the IEEE 13 generator system.

The system consists of,
- Thirteen generators

The data for all these variables are presented in the next section.

5.1.1 Organization of the result

The result are organized under four categories, the values obtained with PSO, APSO, CPSO and NPSO algorithms. After each execution the generator limits of the system are obtained and checked against the limits at the start of the execution.

The best and worst value obtained with each algorithm are compared in the summary of the chapter. The graph presented (no.of iterations Vs total generation cost) were obtained from MATLAB 7. The time of execution was measured after the execution of each algorithm.

5.2 The Test Bus System in Detail

IEEE13 generator system contains the following input datas (for EL problems)

GENERATORS DATA FOR CASE 1 (13 UNITS)

G	P_{min}(MW)	P_{max}(MW)	a	b	c	e	f
1	0	680	0.00028	8.10	550	300	0.035
2	0	360	0.00056	8.10	309	200	0.042
3	0	360	0.00056	8.10	307	200	0.042
4	60	180	0.00324	7.74	240	150	0.063
5	60	180	0.00324	7.74	240	150	0.063
6	60	180	0.00324	7.74	240	150	0.063
7	60	180	0.00324	7.74	240	150	0.063
8	60	180	0.00324	7.74	240	150	0.063
9	60	180	0.00324	7.74	240	150	0.063
10	40	120	0.00284	8.6	126	100	0.084
11	40	120	0.00284	8.6	126	100	0.084
12	55	120	0.00284	8.6	126	100	0.084
13	55	120	0.00284	8.6	126	100	0.084

5.2.(a).Input data for IEEE 13 generator system

5.3 Results obtained by using the PSO

5.3.1 Parameters

1. Total population size =100

2. Total load demand=1800MW

3. Maximum iteration=300

3. Acceleration coefficient c1, c2=2

4. Weight inertia factor,

W_{max}=0.9

W_{min}=0.4

5.3.2 Overall Report

Method	Total Gen.cost($)
PSO(ref)	1.8104e+004
PSO	1.8105e+004

5.3.2.(a).Result obtained by using PSO techniques

Graph of G_{best} versus iterations for best solutions of PSO

The limits for generation cost is fixed the given limits

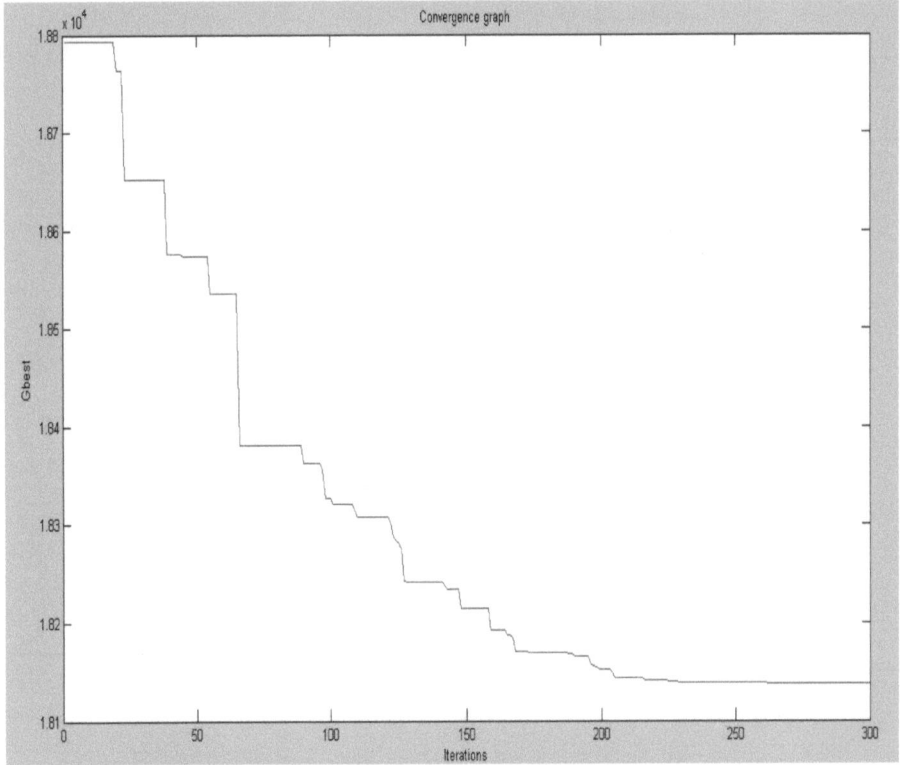

Fig.5.3.2.(b).A graph of Gbest versus iteration for best solutions of PSO

The convergence is achieved at the 220 th iteration.

5.4 Results obtained by using the APSO

5.4.1 Parameters

1. Total population size =100

2. Total load demand=1800 MW

3. Maximum iteration=300

3. Acceleration coefficient c1=0.5, c2=2.5

4. Weight inertia factor,

$$W_{max}=0.9$$

$$W_{min}=0.45$$

5.4.2 Overall Report

Method	Total Gen.cost($)
PSO(ref)	1.8104e+004
PSO	1.8105e+004
SAPSO	1.8076e+004

5.4.2.(a).Result obtained by using APSO techniques

Graph of G$_{best}$ versus iterations for best solutions of APSO

The limits for generation cost is fixed the given limits.

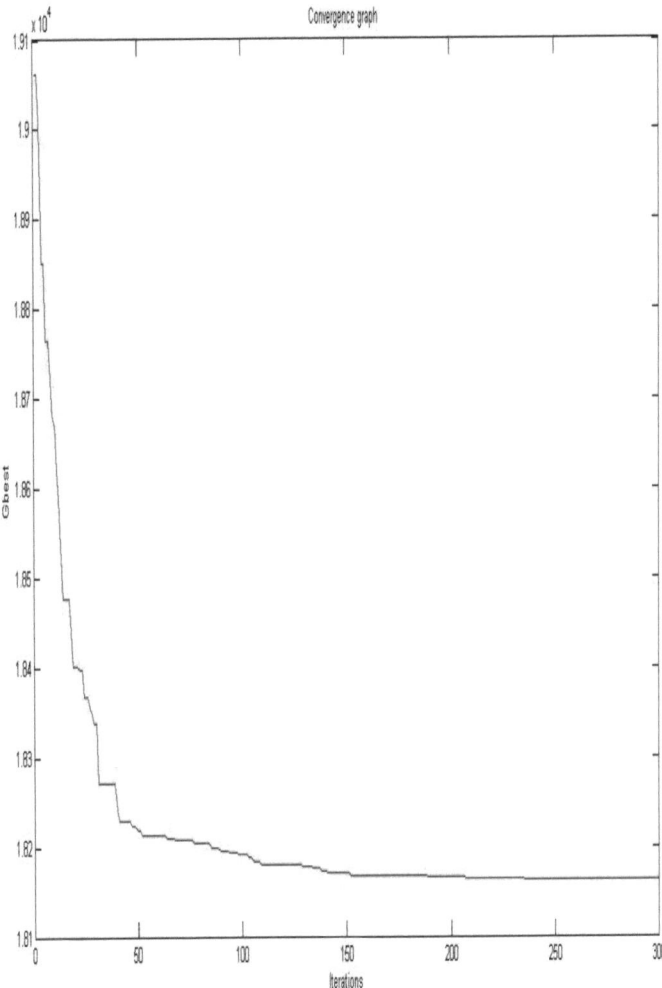

Fig.5.4.2.(b). A graph of Gbest versus iteration for the best solution of APSO
The convergence is achieved at the 150 th iteration.

5.5 Results obtained by using the CPSO

5.5.1 Parameters

1. Total population size =100
2. Total load demand=1800 MW
3. Maximum iteration=300
3. Acceleration coefficient c1, c2=2
4. Weight inertia factor,
   W_{max}=0.9
   W_{min}=0.4

5.5.2 Overall Report

Method	Total Gen.cost($)
PSO(ref)	1.8104e+004
PSO	1.8105e+004
SAPSO	1.8076e+004
CPSO	1.8090e+004

5.5.2.(a).Result obtained by using CPSO techniques

Graph of G_{best} versus iterations for best solutions of CPSO

The limits for generation cost is fixed by the given limits.

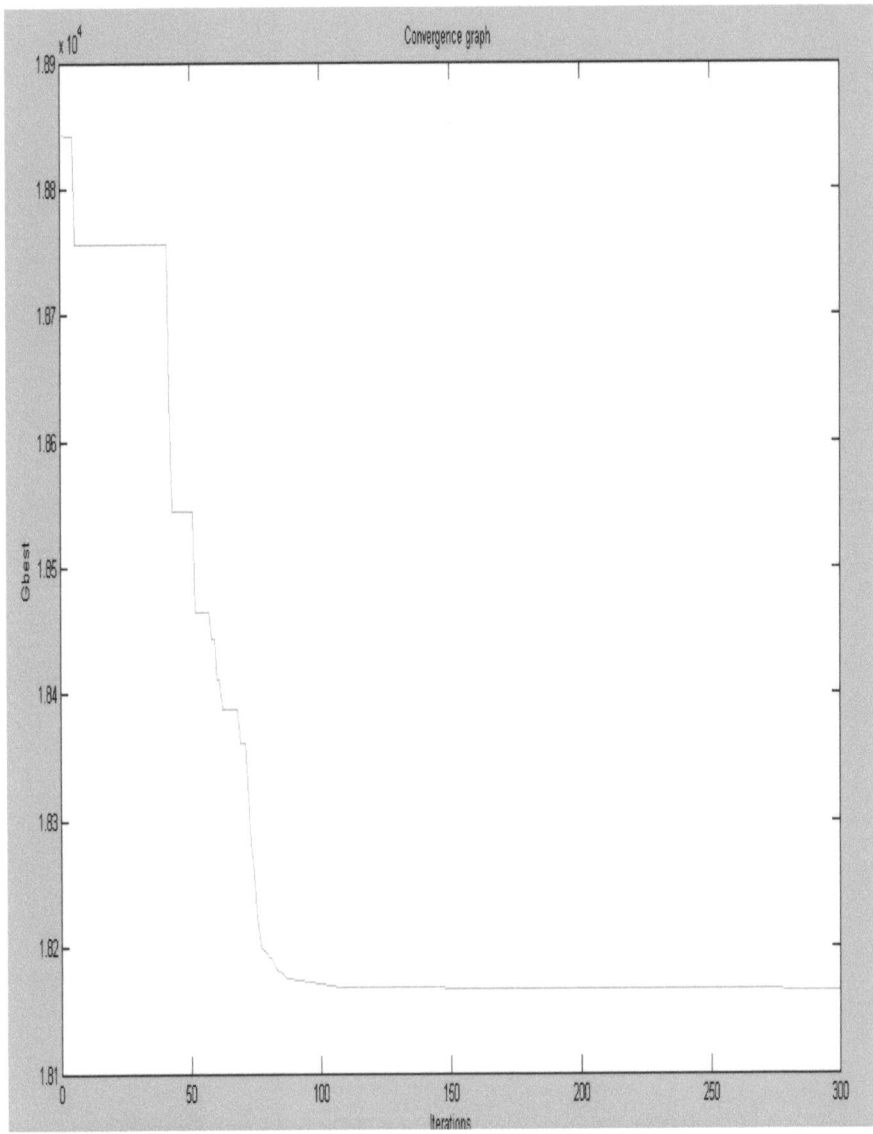

Fig.5.5.2.(b).A graph of Gbest versus iteration for the best solution of CPSO
The convergence is achieved at the 100 th iteration.

5.6 Results obtained by using the NPSO

5.6.1 Parameters

1. Total population size =100
2. Total load demand=1800 MW
3. Maximum iteration=300
3. Acceleration coefficient $c_1=1.4$, $c_2=0.6$
4. Weight inertia factor,
   $W_{max}=0.9$
   $W_{min}=0.4$

5.6.2 Overall Report

Method	Total Gen.cost($)
PSO(ref)	1.8104e+004
PSO	1.8105e+004
SAPSO	1.8076e+004
CPSO	1.8090e+004
NPSO	1.8124e+004

5.6.2.(a).Result obtained by using NPSO techniques

Graph of G_{best} versus iterations for best solutions of NPSO

The limits for generation cost is fixed by the given limits

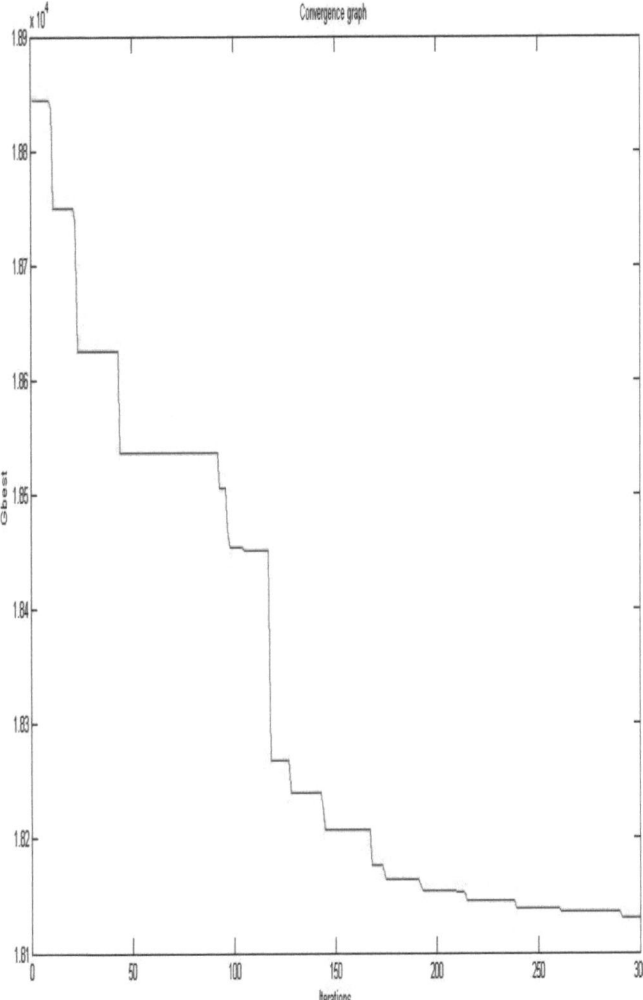

Fig.5.6.2.(b).A graph of Gbest versus iteration for the best solution of NPSO
The convergence is achieved at the 250 th iteration.

5.7 Analysis of four PSO techniques

Algorithm	Gen.cost($)			Time of execution	No. of iterations
	Min.	Max.	Avg.		
PSO(ref)	18014	18207	18014	---------	300
PSO	18029	18178	18015	4.813000	300
CPSO	18074	18101	18090	4.797000	300
APSO	18063	18085	18076	4.954000	300
NPSO	18038	18244	18124	6.125000	300

5.7.(a).Comparison of all four PSO techniques

5.8 Comparison of graphs

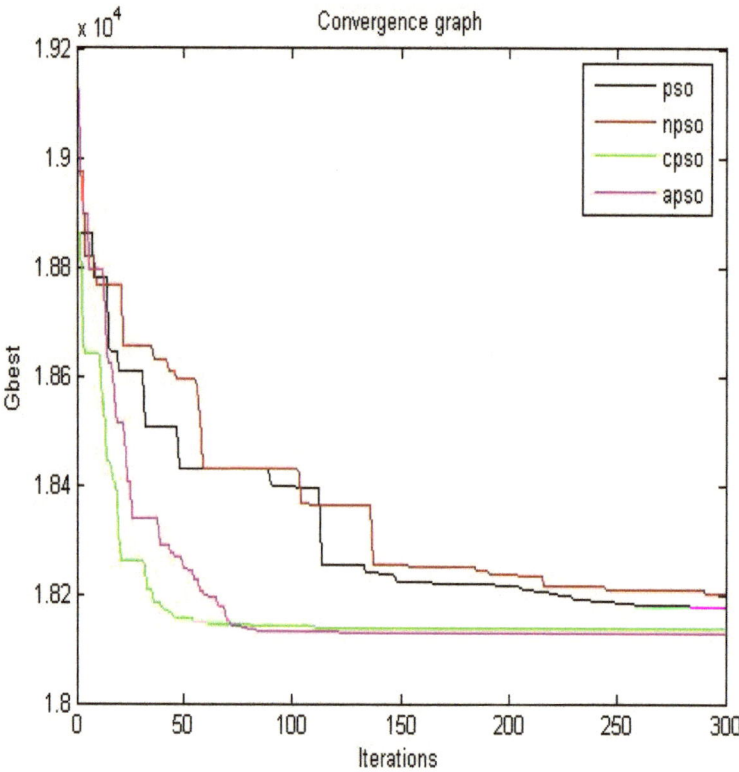

Fig.5.8.(b).Comparison of all four PSO techniques graph

CHAPTER-6

6.1 Analysis of different pso techniques

- From the results it is inferred that Self Adaptive PSO gives the better solution for IEEE 13 generator system when compared to other PSO techniques.
- Speed of convergence is fast in SAPSO and CPSO when compared to other PSO techniques.
- SAPSO and CPSO are robust when compared to other PSO techniques.
- Premature convergence is more in CAPSO when compared to other PSO techniques.
- Computational time is more for New PSO when compared to other PSO techniques.

6.2 Conclusion

Self-Adaptive Particle Swarm Optimization is the most suitable optimization technique for Economic Load Dispatch Problem by comparing various PSO algorithms.

SAMPLE PROGRAMS

[I] PARTICLE SWARM OPTIMIZER

```
clc;
clear;
clf;
% Economic dispatch using PARTICLE SWARM OPTIMIZATION technique

% ----- Defining the limits of the Generators: User Defined-----

N= 100;
%input('Enter the population size of the system : ');
Pmax=[680 360 360 180 180 180 180 180 180 120 120 120 120];
%input('Enter the max generation by individual N generators in the form of an
array : ');
Pmin=[0 0 0 60 60 60 60 60 60 40 40 55 55];
%input('Enter the min generation by individual N generators in the form of an
array : ');
PD=1800;
%input('Give in the total load demand : ');
Pload= [(Pmin)', (Pmax)'];
M=length(Pload);
%the no of generators or independent variables in the system

%-----Take the inertia weight factor limits from the user-----
wmax= .9;
%input('Give the max value of inertia weight factor lying between 0 and 1 : ');
wmin= .4;
%input('Give the min value of inertia weight factor lying between 0 and 1 and
lesser than wmax : ');
maxiter= 300; %input('Enter the maximum number of iterations : ');
R= 10; %input('Enter the number of intervals R : ');
c1= 2; %input ('Input the value of the constant c1 : ');
c2= 2; %input ('Input the value of the constant c2 : ');
tic;

%-----Initializing the population-----
i=1;
while i<=N
    j=1;
    totgen=0;

    while j<=M-1
        P(i,j)=Pload(j,1)+rand*(Pload(j,2)-Pload(j,1));
        totgen=totgen+P(i,j);
        j=j+1;
    end

    P(i,j)=PD-totgen;

    if(P(i,j)>=(Pload(M,1)&&P(i,j)<=Pload(M,2)))
        i=i+1;
    end

end

%-----Initializing the velocity limits-----
```

```
for i=1:M
    Vlimit(i,1)= (Pload(i,1)-Pload(i,2))/R;
    Vlimit(i,2)= -Vlimit(i,1);
end

%-----Generating the initial velocities-----

for i= 1 :N
    for j=1:M-1
        V(i,j)=( Vlimit(j,1)+rand*( Vlimit(j,2)-Vlimit(j,1) ));
    end
end


%-----Starting the iterations-----

Pbest=P;
%initializing the matrix Pbest

for iter=1:maxiter
%loop for iterations starts here

for i= 1 :N
%finding out fitness of Pbest
        for j=1:M
        gen(j)=Pbest(i,j);
        end
        Pfit(i)=fit_thirteen(gen,Pmin);
    end

%-----Finding out Gbest-----

    Gfit=min(Pfit);
    best_pos=find(Pfit==min(Pfit));
    Gbest=P(best_pos(1),:);

    W=wmax-((wmax-wmin)*iter/maxiter);
%intertia constant
    for i=1:N
        for j=1:M-1
            V(i,j)=(V(i,j)*W)+( c1*rand*(Pbest(i,j)-P(i,j)) ) + (
c2*rand*(Gbest(j)-P(i,j)) );
            if V(i,j)<Vlimit(j,1)
                V(i,j)=Vlimit(j,1);
            end
            if V(i,j)>Vlimit(j,2)
                V(i,j)=Vlimit(j,2);
            end
        end
    end
%-----updating the population-----
    for i=1:N
        sum=0;

        for j=1:M-1
            P(i,j)=P(i,j)+V(i,j);

            if P(i,j)<Pload(j,1)
                P(i,j)=Pload(j,1);
            end
            if  P(i,j)>Pload(j,2)
```

```matlab
                P(i,j)=Pload(j,2);
            end

            sum=sum+P(i,j);
        end

        P(i,M)=PD-sum;
        fit(i)=0;

        if P(i,M)<Pload(M,1)
            pen=Pload(M,1)-P(i,M);
            fit(i)=100*(pen)^2;
        end

        if P(i,M)>Pload(M,2)
            pen=P(i,M)-Pload(M,2);
                fit(i)=100*(pen)^2;
        end
    end

    %-----Finding out the fitness of the present population-----
    for i= 1 :N
        for j=1:M
        gen(j)=P(i,j);
        end
        fit(i)=fit(i)+fit_thirteen(gen,Pmin);
    end
    for i=1:N
        if(fit(i)<Pfit(i))
            Pbest(i,:)=P(i,:);
            best_pos=find(fit==min(fit));
            Gbest=P(best_pos(1),:);
        end
    end
    conver(iter)=min(Pfit);
%for plotting convergence
end
fprintf('\nThe cost using PSO was found out to be: ');
Gfit
fprintf('\nThe best generation using PSO was found out to be: ');
Gbest

toc;

plot(conver,'r');
xlabel('Iterations');
ylabel('Gbest');
title('Convergence graph');
%hold on
%NPSO_13gen


% -------------GOOD OUTPUT----------------
% Give in the total load demand : 1800
%
% The cost using PSO was found out to be:
% Gfit =
%
%    1.8076e+004
%
%
% The best generation using PSO was found out to be:
% Gbest =
```

```
%    628.3185  299.1993  299.1993   60.0000   60.0000   60.0000   60.0000
% 60.0000   60.0000   40.0000   40.0000   55.0000   78.2829
%
% Elapsed time is 6.031000 seconds.
% Give in the total load demand : 1800
%
% The cost using NPSO was found out to be:
% Gfit =
%
%    1.8202e+004
%
%
% The best generation using NPSO was found out to be:
% Gbest =
%
%    628.7357  360.0000  221.8877   60.0000   60.0000   60.0000   60.0000
% 60.0000   60.0000   40.0000   40.0000   55.0000   94.3766
%
% Elapsed time is 6.562000 seconds.
% Give in the total load demand : 1800
%
% The cost using DPSO was found out to be:
% Gfit =
%
%    1.7989e+004
%
%
% The best generation using DPSO was found out to be:
% Gbest =
%
%    628.2928  223.5379  298.4599   60.0000   60.0000   60.0000  159.7263
% 60.0000   60.0000   40.0000   40.0003   55.0000   54.9828
%
% Elapsed time is 7.625000 seconds.
```

[II] ADAPTIVE PARTICLE SWARM OPTIMIZER

```
clc;
clear;
clf;
% Economic dispatch using PARTICLE SWARM OPTIMIZATION technique

% ----- Defining the limits of the Generators: User Defined-----

N= 100; %input('Enter the population size of the system : ');
Pmax=[680 360 360 180 180 180 180 180 180 120 120 120 120];
%input('Enter the max generation by individual N generators in the form of an
array : ');
Pmin=[0 0 0 60 60 60 60 60 60 40 40 55 55];
%input('Enter the min generation by individual N generators in the form of an
array : ');
PD=1800;
%input('Give in the total load demand : ');

Pload= [(Pmin)', (Pmax)'];
M=length(Pload);
%the no of generators or independent variables in the system

%-----Take the inertia weight factor limits from the user-----
wmax= .9;
%input('Give the max value of inertia weight factor lying between 0 and 1 : ');
wmin= .4;
%input('Give the min value of inertia weight factor lying between 0 and 1 and
lesser than wmax : ');

maxiter= 300;
%input('Enter the maximum number of iterations : ');

R= 10;
%input('Enter the number of intervals R : ');

c1= 2;
%input ('Input the value of the constant c1 : ');
c2= 2;
%input ('Input the value of the constant c2 : ');
c1=2.05;
cf=0.5;
ce=2.5;
wmax=0.9;
wmin=0.45;
tic;
conver1=zeros(1,300);

%-----Initializing the population-----
i=1;
while i<=N
    j=1;
    totgen=0;

    while j<=M-1
        P(i,j)=Pload(j,1)+rand*(Pload(j,2)-Pload(j,1));
        totgen=totgen+P(i,j);
        j=j+1;
    end
```

```
        P(i,j)=PD-totgen;

        if(P(i,j)>=(Pload(M,1)&&P(i,j)<=Pload(M,2)))
            i=i+1;
        end

    end

%-----Initializing the velocity limits-----

for i=1:M
    Vlimit(i,1)= (Pload(i,1)-Pload(i,2))/R;
    Vlimit(i,2)= -Vlimit(i,1);
end

%-----Generating the initial velocities-----

for i= 1 :N
    for j=1:M-1
        V(i,j)=( Vlimit(j,1)+rand*( Vlimit(j,2)-Vlimit(j,1) ));
    end
end


%-----Starting the iterations-----

Pbest=P;
%initializing the matrix Pbest

for iter=1:maxiter
%loop for iterations starts here

    for i= 1 :N
%finding out fitness of Pbest
        for j=1:M
        gen(j)=Pbest(i,j);
        end

        Pfit(i)=fit_thirteen1(gen,Pmin);
        fita(i)=Pfit(i);

    end

    %-----Finding out Gbest-----

    Gfit=min(Pfit);
    best_pos=find(Pfit==min(Pfit));
    Gbest=P(best_pos(1),:);

% W=wmax-((wmax-wmin)*iter/maxiter);
%intertia constant
    for i=1:N
        geta=(fita(i)-Gfit)/fita(i);
        fgeta=2*(1-cos((pi*geta)/2));
        W=(wmax*fgeta)+wmin;
        ggeta=2.5*(1-cos((pi*geta)/2));
        c2=(cf*ggeta)+ce;
        for j=1:M-1
            V(i,j)=(V(i,j)*W)+( c1*rand*(Pbest(i,j)-P(i,j)) ) + ( c2*rand*(Gbest(j)-P(i,j)) );
```

```matlab
                if V(i,j)<Vlimit(j,1)
                    V(i,j)=Vlimit(j,1);
                end
                if V(i,j)>Vlimit(j,2)
                    V(i,j)=Vlimit(j,2);
                end
            end
     end
%-----updating the population-----
    for i=1:N
        sum=0;

        for j=1:M-1
            P(i,j)=P(i,j)+V(i,j);

            if P(i,j)<Pload(j,1)
                P(i,j)=Pload(j,1);
            end
            if  P(i,j)>Pload(j,2)
                P(i,j)=Pload(j,2);
            end

            sum=sum+P(i,j);
        end

        P(i,M)=PD-sum;
        fita(i)=0;

        if P(i,M)<Pload(M,1)
            pen=Pload(M,1)-P(i,M);
            fita(i)=100*(pen)^2;
        end

        if P(i,M)>Pload(M,2)
            pen=P(i,M)-Pload(M,2);
                fita(i)=100*(pen)^2;
        end
    end

 %-----Finding out the fitness of the present population-----
    for i= 1 :N
        for j=1:M
        gen(j)=P(i,j);
        end
        fita(i)=fita(i)+fit_thirteen1(gen,Pmin);
    end
    for i=1:N
        if(fita(i)<Pfit(i))
            Pbest(i,:)=P(i,:);
            best_pos=find(fita==min(fita));
            Gbest=P(best_pos(1),:);
        end
    end
    conver1(iter)=min(Pfit);
%for plotting convergence
end
fprintf('\nThe cost using APSO was found out to be: ');
Gfit
fprintf('\nThe best generation using APSO was found out to be: ');
Gbest

toc;
```

```
plot(conver1,'r')
xlabel('Iterations');
ylabel('Gbest');
title('Convergence graph');

% -------------GOOD OUTPUT----------------
% Give in the total load demand : 1800
%
% The cost using PSO was found out to be:
% Gfit =
%
%    1.8076e+004
%
%
% The best generation using PSO was found out to be:
% Gbest =
%
%   628.3185  299.1993  299.1993   60.0000   60.0000   60.0000   60.0000
%    60.0000   60.0000   40.0000   40.0000   55.0000   78.2829
%
% Elapsed time is 6.031000 seconds.
% Give in the total load demand : 1800
%
% The cost using NPSO was found out to be:
% Gfit =
%
%    1.8202e+004
%
%
% The best generation using NPSO was found out to be:
% Gbest =
%
%   628.7357  360.0000  221.8877   60.0000   60.0000   60.0000   60.0000
%    60.0000   60.0000   40.0000   40.0000   55.0000   94.3766
%
% Elapsed time is 6.562000 seconds.
% Give in the total load demand : 1800
%
% The cost using DPSO was found out to be:
% Gfit =
%
%    1.7989e+004
%
%
% The best generation using DPSO was found out to be:
% Gbest =
%
%   628.2928  223.5379  298.4599   60.0000   60.0000   60.0000  159.7263
%    60.0000   60.0000   40.0000   40.0003   55.0000   54.9828
%
% Elapsed time is 7.625000 seconds.
```

[III] CHAOTIC PARTICLE SWARM OPTIMIZER

```
% clc;
% clear;
% clf;
% Economic dispatch using PARTICLE SWARM OPTIMIZATION technique

% ----- Defining the limits of the Generators: User Defined-----

N= 100;
%input('Enter the population size of the system : ');
Pmax=[680 360 360 180 180 180 180 180 180 120 120 120 120];
%input('Enter the max generation by individual N generators in the form of an
array : ');
Pmin=[0 0 0 60 60 60 60 60 60 40 40 55 55];
%input('Enter the min generation by individual N generators in the form of an
array : ');
PD=1800;
%input('Give in the total load demand : ');

Pload= [(Pmin)', (Pmax)'];
M=length(Pload);
%the no of generators or independent variables in the system

%-----Take the inertia weight factor limits from the user-----
wmax= .9;
%input('Give the max value of inertia weight factor lying between 0 and 1 : ');
wmin= .4;
%input('Give the min value of inertia weight factor lying between 0 and 1 and
lesser than wmax : ');

maxiter= 300;
%input('Enter the maximum number of iterations : ');

R= 10;
%input('Enter the number of intervals R : ');

c1= 2;
%input ('Input the value of the constant c1 : ');
c2= 2;
%input ('Input the value of the constant c2 : ');
W=0.48;
tic;

%-----Initializing the population-----
i=1;
while i<=N
    j=1;
    totgen=0;

    while j<=M-1
        P(i,j)=Pload(j,1)+rand*(Pload(j,2)-Pload(j,1));
        totgen=totgen+P(i,j);
        j=j+1;
    end

    P(i,j)=PD-totgen;

    if(P(i,j)>=(Pload(M,1)&&P(i,j)<=Pload(M,2)))
        i=i+1;
```

```
        end

end

%-----Initializing the velocity limits-----

for i=1:M
    Vlimit(i,1)= (Pload(i,1)-Pload(i,2))/R;
    Vlimit(i,2)= -Vlimit(i,1);
end

%-----Generating the initial velocities-----

for i= 1 :N
   for j=1:M-1
      V(i,j)=( Vlimit(j,1)+rand*( Vlimit(j,2)-Vlimit(j,1) ));
   end
end


%-----Starting the iterations-----

Pbest=P;
%initializing the matrix Pbest

for iter=1:maxiter
%loop for iterations starts here

    for i= 1 :N
%finding out fitness of Pbest
        for j=1:M
        gen(j)=Pbest(i,j);
        end
        Pfit(i)=fit_thirteen(gen,Pmin);
    end

    %-----Finding out Gbest-----

    Gfit=min(Pfit);
    best_pos=find(Pfit==min(Pfit));
    Gbest=P(best_pos(1),:);

    W=4*W*(1-W);
%intertia constant
    for i=1:N
        for j=1:M-1
            V(i,j)=(V(i,j)*W)+( c1*rand*(Pbest(i,j)-P(i,j)) ) + ( c2*rand*(Gbest(j)-P(i,j)) );
            if V(i,j)<Vlimit(j,1)
                V(i,j)=Vlimit(j,1);
            end
            if V(i,j)>Vlimit(j,2)
                V(i,j)=Vlimit(j,2);
            end
        end
    end
%-----updating the population-----
   for i=1:N
       sum=0;

       for j=1:M-1
```

```
            P(i,j)=P(i,j)+V(i,j);

            if P(i,j)<Pload(j,1)
               P(i,j)=Pload(j,1);
            end
            if  P(i,j)>Pload(j,2)
                P(i,j)=Pload(j,2);
            end

            sum=sum+P(i,j);
        end

        P(i,M)=PD-sum;
        fit(i)=0;

        if P(i,M)<Pload(M,1)
           pen=Pload(M,1)-P(i,M);
           fit(i)=100*(pen)^2;
        end

        if P(i,M)>Pload(M,2)
           pen=P(i,M)-Pload(M,2);
                fit(i)=100*(pen)^2;
        end
    end

    %-----Finding out the fitness of the present population-----
    for i= 1 :N
        for j=1:M
        gen(j)=P(i,j);
        end
        fit(i)=fit(i)+fit_thirteen(gen,Pmin);
    end
    for i=1:N
        if(fit(i)<Pfit(i))
           Pbest(i,:)=P(i,:);
           best_pos=find(fit==min(fit));
           Gbest=P(best_pos(1),:);
        end
    end
    conver(iter)=min(Pfit);
%for plotting convergence
end
fprintf('\nThe cost using CPSO was found out to be: ');
Gfit
fprintf('\nThe best generation using CPSO was found out to be: ');
Gbest

toc;

plot(conver,'g');
xlabel('Iterations');
ylabel('Gbest');
title('Convergence graph');
%hold on
%NPSO_13gen


% --------------GOOD OUTPUT----------------
% Give in the total load demand : 1800
%
% The cost using PSO was found out to be:
% Gfit =
```

```
%    1.8076e+004
%
%
% The best generation using PSO was found out to be:
% Gbest =
%
%   628.3185  299.1993  299.1993   60.0000   60.0000   60.0000   60.0000
%    60.0000   60.0000   40.0000   40.0000   55.0000   78.2829
%
% Elapsed time is 6.031000 seconds.
% Give in the total load demand : 1800
%
% The cost using NPSO was found out to be:
% Gfit =
%
%    1.8202e+004
%
%
% The best generation using NPSO was found out to be:
% Gbest =
%
%   628.7357  360.0000  221.8877   60.0000   60.0000   60.0000   60.0000
%    60.0000   60.0000   40.0000   40.0000   55.0000   94.3766
%
% Elapsed time is 6.562000 seconds.
% Give in the total load demand : 1800
%
% The cost using DPSO was found out to be:
% Gfit =
%
%    1.7989e+004
%
%
% The best generation using DPSO was found out to be:
% Gbest =
%
%   628.2928  223.5379  298.4599   60.0000   60.0000   60.0000  159.7263
%    60.0000   60.0000   40.0000   40.0003   55.0000   54.9828
%
% Elapsed time is 7.625000 seconds.
```

[IV] NEW PARTICLE SWARM OPTIMIZER

```
clc;
clear;
clf;
% Economic dispatch using PARTICLE SWARM OPTIMIZATION technique

% ----- Defining the limits of the Generators: User Defined-----

N= 100;
%input('Enter the population size of the system : ');
Pmax=[680 360 360 180 180 180 180 180 180 120 120 120 120];
%input('Enter the max generation by individual N generators in the form of an
array : ');
Pmin=[0 0 0 60 60 60 60 60 60 40 40 55 55];
%input('Enter the min generation by individual N generators in the form of an
array : ');
PD=1800;
%input('Give in the total load demand : ');

Pload= [(Pmin)', (Pmax)'];
M=length(Pload);
%the no of generators or independent variables in the system

%-----Take the inertia weight factor limits from the user-----
wmax= .9;
%input('Give the max value of inertia weight factor lying between 0 and 1 : ');
wmin= .4;
%input('Give the min value of inertia weight factor lying between 0 and 1 and
lesser than wmax : ');

maxiter= 300;
%input('Enter the maximum number of iterations : ');

R= 10;
%input('Enter the number of intervals R : ');

c1b= 1.6;
%input ('Input the value of the constant c1 : ');
c1w=0.4;
c2= 2;
%input ('Input the value of the constant c2 : ');

tic;
conver1=zeros(1,300);

%-----Initializing the population-----
i=1;
while i<=N
    j=1;
    totgen=0;

    while j<=M-1
        P(i,j)=Pload(j,1)+rand*(Pload(j,2)-Pload(j,1));
        totgen=totgen+P(i,j);
        j=j+1;
    end

    P(i,j)=PD-totgen;
```

```matlab
        if(P(i,j)>=(Pload(M,1)&&P(i,j)<=Pload(M,2)))
            i=i+1;
        end
    end
end

%-----Initializing the velocity limits-----

for i=1:M
    Vlimit(i,1)= (Pload(i,1)-Pload(i,2))/R;
    Vlimit(i,2)= -Vlimit(i,1);
end

%-----Generating the initial velocities-----

for i= 1 :N
    for j=1:M-1
        V(i,j)=( Vlimit(j,1)+rand*( Vlimit(j,2)-Vlimit(j,1) ));
    end
end


%-----Starting the iterations-----

Pbest=P;
%initializing the matrix Pbest
Pworst=P;
for iter=1:maxiter
%loop for iterations starts here

    for i= 1 :N
%finding out fitness of Pbest
        for j=1:M
            gen(j)=Pbest(i,j);
        end
        Pfit(i)=fit_thirteen(gen,Pmin);
    end
    for i= 1 :N
%finding out fitness of worst
        for j=1:M
            gen(j)=Pworst(i,j);
        end
        Pwfit(i)=fit_thirteen(gen,Pmin);
    end

    %-----Finding out Gbest-----

    Gfit=min(Pfit);
    best_pos=find(Pfit==min(Pfit));
    Gbest=P(best_pos(1),:);

    W=wmax-((wmax-wmin)*iter/maxiter);
%intertia constant
    for i=1:N
        for j=1:M-1
            V(i,j)=(V(i,j)*W)+( c1b*rand*(Pbest(i,j)-P(i,j)) ) + ( c2*rand*(Gbest(j)-P(i,j)) )-( c1w*rand*(Pworst(i,j)-P(i,j)) );
            if V(i,j)<Vlimit(j,1)
                V(i,j)=Vlimit(j,1);
            end
            if V(i,j)>Vlimit(j,2)
```

```
                    V(i,j)=Vlimit(j,2);
            end
        end
    end
%-----updating the population-----
    for i=1:N
        sum=0;

        for j=1:M-1
            P(i,j)=P(i,j)+V(i,j);

            if P(i,j)<Pload(j,1)
               P(i,j)=Pload(j,1);
            end
            if  P(i,j)>Pload(j,2)
                P(i,j)=Pload(j,2);
            end

            sum=sum+P(i,j);
        end

        P(i,M)=PD-sum;
        fit(i)=0;

        if P(i,M)<Pload(M,1)
           pen=Pload(M,1)-P(i,M);
           fit(i)=100*(pen)^2;
        end

        if P(i,M)>Pload(M,2)
           pen=P(i,M)-Pload(M,2);
                fit(i)=100*(pen)^2;
        end
    end

%-----Finding out the fitness of the present population-----
    for i= 1 :N
        for j=1:M
        gen(j)=P(i,j);
        end
        fit(i)=fit(i)+fit_thirteen(gen,Pmin);
    end
    for i=1:N
        if(fit(i)<Pfit(i))
           Pbest(i,:)=P(i,:);
           best_pos=find(fit==min(fit));
           Gbest=P(best_pos(1),:);
        end
    end
    for i=1:N
        if(fit(i)>Pwfit(i))
           Pworst(i,:)=P(i,:);
                end
    end
    conver1(iter)=min(Pfit);
%for plotting convergence
end
fprintf('\nThe cost using NPSO was found out to be: ');
Gfit
fprintf('\nThe best generation using NPSO was found out to be: ');
Gbest
```

```
toc;

plot(conver1,'r')
xlabel('Iterations');
ylabel('Gbest');
title('Convergence graph');

% -------------GOOD OUTPUT----------------
% Give in the total load demand : 1800
%
% The cost using PSO was found out to be:
% Gfit =
%
%    1.8076e+004
%
%
% The best generation using PSO was found out to be:
% Gbest =
%
%   628.3185  299.1993  299.1993   60.0000   60.0000   60.0000   60.0000
%  60.0000   60.0000   40.0000   40.0000   55.0000   78.2829
%
% Elapsed time is 6.031000 seconds.
% Give in the total load demand : 1800
%
% The cost using NPSO was found out to be:
% Gfit =
%
%    1.8202e+004
%
%
% The best generation using NPSO was found out to be:
% Gbest =
%
%   628.7357  360.0000  221.8877   60.0000   60.0000   60.0000   60.0000
%  60.0000   60.0000   40.0000   40.0000   55.0000   94.3766
%
% Elapsed time is 6.562000 seconds.
% Give in the total load demand : 1800
%
% The cost using DPSO was found out to be:
% Gfit =
%
%    1.7989e+004
%
%
% The best generation using DPSO was found out to be:
% Gbest =
%
%   628.2928  223.5379  298.4599   60.0000   60.0000   60.0000  159.7263
%  60.0000   60.0000   40.0000   40.0003   55.0000   54.9828
%
% Elapsed time is 7.625000 seconds.
```

REFERENCES

[1] C.H.Chen, S.N.Yeh senior member IEEE, **"Particle Swarm Optimization for Economic Power Dispatch with Valve-Point Effect"**.

[2] M.Sudhakaran, P.Ajay-D-Vimal Raj and T.G.Palanivelu: **"Application of Particle Swarm Optimization for Economic Load Dispatch Problems"**.

[3] Zhengjia Wu, Jianzhong Zhou **"A Self-adaptive Particle Swarm Optimization Algorithm with Individual Coefficients Adjustment"**.

[4] D.N.Jeyakumar, T.Jayabarathi, T.Raghunathan: **"Particle Swarm Optimization for various types of Economic Dispatch problems"**.

[5] Nidul Sinha, R.Chakrabarti, and P.K.Chattopadhyay,: **"Evolutionary Programming Techniques for Economic Load Dispatch"**.

[6] Cai Jiejin, Ma Xiaoqian, Li Lixiang, Peng Haipeng, **"Chaotic Particle Swarm Optimization for Economic Dispatch Problems"**.

[7] A. Immanuel Selvakumar, K. Thanushkodi **"A New Particle Swarm Optimization Solution to Nonconvex Economic Dispatch Problems"**.

[8] Allen J.Wood, Bruce F.Wollenberg **"Power generation, operation and Control"**.
Haadi Sadat **"Power System"**, presents an excellent introduction to the Economic Dispatch problems.

[9] Zwe-Lee Gaing **"Particle Swarm Optimization to solving the Economic Dispatch Considering the Generators Constraints"**.